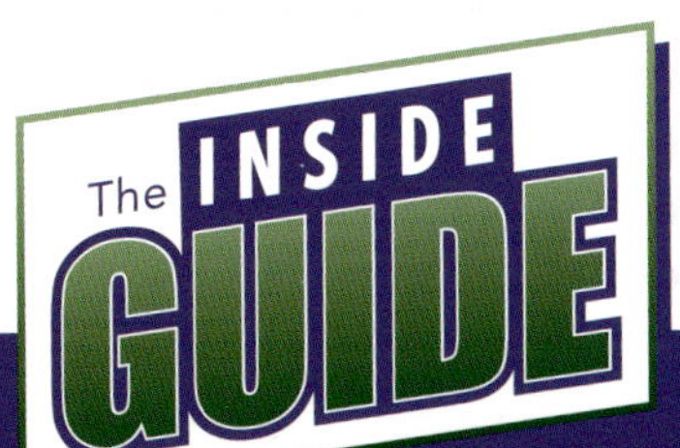

ROBOTICS

Robots in Space

By Spencer Parks

New York

Published in 2022 by Cavendish Square Publishing, LLC
243 5th Avenue, Suite 136, New York, NY 10016

First Edition

Website: cavendishsq.com

Portions of this work were originally authored by Ryan Nagelhout and published as *Space Robots (Robots and Robotics)*. All new material this edition authored by Spencer Parks.

All websites were available and accurate when this book was sent to press.

Library of Congress Cataloging-in-Publication Data

Names: Parks, Spencer, author.
Title: Robots in space / Spencer Parks.
Description: New York : Cavendish Square Publishing, 2022. | Series: The inside guide : robotics | Includes bibliographical references and index.
Identifiers: LCCN 2020027226 | ISBN 9781502660688 (library binding) | ISBN 9781502660664 (paperback) | ISBN 9781502660671 (set) | ISBN 9781502660695 (ebook)
Subjects: LCSH: Space robotics–Juvenile literature.
Classification: LCC TL1097 .P374 2022 | DDC 629.47–dc23
LC record available at https://lccn.loc.gov/2020027226

Editor: Caitie McAneney
Copyeditor: Jill Keppeler
Designer: Deanna Paternostro

The photographs in this book are used by permission and through the courtesy of: Cover, pp. 4, 6 (inset), 8 (bottom), 18 (bottom) 19 (top and bottom), 24 Courtesy of NASA; p. 6 (main) John B. Carnett/Contributor/Popular Science/Getty Images; p. 8 (top) Sovfoto/Contributor/Universal Images Group/Getty Images; p. 9 (background) NASA, ESA, J. Hester, A. Loll (ASU); pp. 9 (inset), 16 NASA/Handout/Getty Images News/Getty Images; p. 10 Space Frontiers/Stringer/Archive Photos/Getty Images; pp. 11, 12 (bottom) NASA/JPL-Caltech/MSSS; p. 12 (top) NASA/JPL/Cornell; p. 13 Historical/Contributor/Corbis Historical/Getty Images; p. 14 (top) Science & Society Picture Library/Contributor/SSPL/Getty Images; p. 14 (center) Chip Somodevilla/Staff/Getty Images News/Getty Images; p. 14 (bottom) AKSARAN/Contributor/Gamma-Rapho/Getty Images; p. 15 Time & Life Pictures/The LIFE Picture Collection/Getty Images; p. 18 (top) The Washington Post/Contributor/The Washington Post/Getty Images; p. 21 NASA/Anne McClain; p. 22 Bettmann/Contributor/Bettmann/Getty Images; p. 25 NASA/JPL/Space Science Institute; p. 26 MARK RALSTON/Staff/AFP/Getty Images; p. 27 m-gucci/iStock/Getty Images Plus/Getty Images; p. 28 (top) VCG/Contributor/Visual China Group/Getty Images; p. 28 (bottom) MIKE NELSON/Staff/AFP/Getty Images; p. 29 (top) picture alliance/Contributor/picture alliance/Getty Images; p. 29 (bottom) KIRILL KUDRYAVTSEV/Staff/AFP/Getty Images.

CPSIA compliance information: Batch #CS22CSQ: For further information contact Cavendish Square Publishing LLC, New York, New York, at 1-877-980-4450.

Printed in the United States of America

CONTENTS

People have lived on the International Space Station continuously since November 2000. The ISS houses six people at a time, with six bedrooms, two bathrooms, and a gym.

ROBOTS OUT OF THIS WORLD

The 20th century saw both the invention of true robots and the first explorations of outer space. In fact, robots helped make it possible for humans to discover—and travel to—places outside Earth's **atmosphere**. Today, we have many kinds of robots in space. Some are traveling millions—even billions—of miles away, while others help the people who live and work in space aboard the International Space Station (ISS).

Built for Space!

Robots are often used to help people perform dangerous or **repetitive** jobs. They come in different shapes and sizes, and each robot is built for the job it's meant to do. However, most robots share the same basic parts. Sensors help a robot gather information about its surroundings. Effectors, such as robotic arms, help the robot interact with its surroundings. Actuators are the motors that move the parts of a robot, such as the effectors. The controller is a robot's "brain," guiding the robot through preprogrammed actions. Other robots are remote controlled (RC), or controlled from a distance.

Fast Fact

Most space robots leave Earth carried by rockets, which use powerful fuel to propel, or move, the robot beyond the pull of Earth's gravity and into space.

Space robots need special parts and materials so they can function outside Earth. Once they leave Earth, many robots are powered by solar panels, which collect energy from the sun. The panels turn the sunlight into power, which allows the robot to move and communicate. Some robots are built to work alongside human operators, while others are sent out on long space journeys alone.

Fast Fact

The Deep Space Network has three major facilities that send signals to and receive signals from robots in space. They are in the Mojave Desert of California in the United States; near Madrid, Spain; and near Canberra, Australia.

The largest antennas in the Deep Space Network are around 230 feet (70 meters) across.

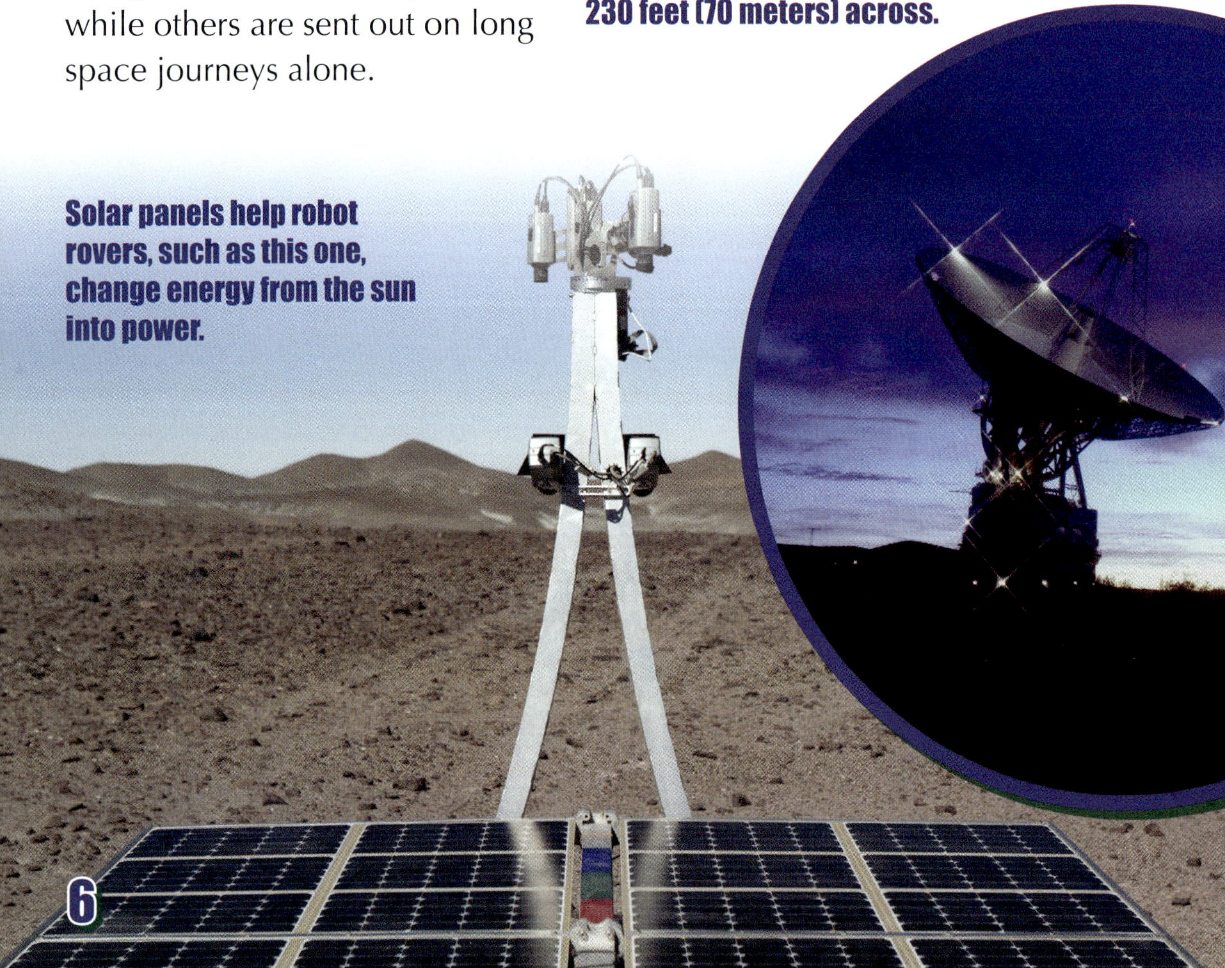

Solar panels help robot rovers, such as this one, change energy from the sun into power.

ROBOT COMMUNICATION

Some robots use **artificial intelligence** to work autonomously, or on their own. However, most robots are guided in some way by a human operator. How can operators on Earth guide robots in space? Robots communicate with scientists on Earth using antennas, or devices that send messages back to our planet. Many space robots are controlled from an international network of antennas called the Deep Space Network (DSN). Sometimes, robots on Mars send signals to a robot in **orbit**, which sends signals back to Earth, and the other way around.

The First Spacecraft

The first Earthlings to go to space weren't humans—they were **satellites**. On October 4, 1957, the Soviet Union launched the first man-made satellite, called Sputnik, into space. Just two years later, the Soviet Union sent the *Luna 2* to the moon's surface. NASA (the U.S. National Aeronautics and Space Administration) sent the *Mariner 2* to Venus in 1962 and the *Mariner 9* to orbit Mars in 1971. In 1997, NASA sent the *Sojourner* rover in the Mars *Pathfinder* spacecraft to explore the surface of Mars.

One of the biggest breakthroughs in spacecraft history was the building of the ISS. In 2000, the first crew arrived to live on the orbiting station. More than 240 astronauts from around the world have lived and worked on the ISS, and robotics makes it all possible. Astronauts perform important experiments in different areas of the station, and it takes over 50 computers to keep the ISS running.

Sputnik was only about the size of a beach ball, but it kicked off the Space Race between the United States and the Soviet Union.

Picture This

You might look up at the stars and moon and wonder what bodies and events in space really look like. Thanks to satellites, rovers, and telescopes, we have many images from our solar system and beyond.

Launched in 1990, the Hubble Space Telescope is one of the most advanced telescopes ever built. It focuses its camera **lens** deep into space, sending hundreds of thousands of amazing images back to Earth.

While some may not consider telescopes to be robots, the Hubble Space Telescope uses sensors, a power source, and scientific instruments to do its job—take photographs of deep space.

Astronauts use this robotic arm from a space shuttle to fix the Hubble Space Telescope.

This school bus–sized telescope has made more than 1.3 million observations and is one of the most productive scientific instruments ever created. ISS astronauts use a robotic arm to make repairs to the giant telescope.

Hubble has studied black holes, space objects, and exciting parts of deep space. It's helped scientists learn the age of our universe and find new planets far outside our solar system.

Fast Fact

Hubble is such a sensitive and powerful telescope that it can detect a night-light being used on the moon's surface.

The first Mars rover, *Sojourner*, landed on Mars in 1997. It stayed within 40 feet (12.2 m) of its lander and took more than 500 photographs.

Chapter Two

ROVERS AT WORK

Humans haven't made it to Mars yet—but robots have, and they're on a mission. From the first Mars rover to today's *Curiosity*, this kind of robot has unlocked many secrets of Mars.

Robot Reconnaissance

Reconnaissance means finding information about something. Because humans can't travel very far in space, robots are our main source of information. Today, the robots we use to study space fall into three basic groups: orbiters, landers, and rovers.

Orbiters are robots designed to orbit an object such as a planet or moon. Landers are designed to reach the surface of a planet, a moon, or another body in space. Rovers are designed to reach the surface of a body in space and travel to certain places on its surface.

***Curiosity* took this selfie on Mount Sharp on Mars.**

Fast Fact

Unfortunately, Mars can be rough on even the toughest rover! Scientists lost communication with the *Spirit* rover in 2010 and the *Opportunity* rover in 2018.

Mars has some violent dust storms. *Opportunity*'s mission was officially ended in 2019, after communication was lost in 2018 following a dust storm on Mars.

The space rovers *Spirit*, *Opportunity*, and *Curiosity* have traveled across many regions of Mars for reconnaissance missions. They've collected mineral samples, taken pictures, and used special sensors to track weather events. The *Perseverance* rover landed on Mars in February 2021 with big plans to explore the Red Planet.

Curious About Mars

NASA launched the *Curiosity* rover on November 26, 2011, and it landed on Mars on August 6, 2012. As of early 2021, this car-sized rover is still in use. *Curiosity* was designed to study the climate and geography of Mars. It also searches for signs of carbon—an element that's considered one of the building blocks of life. The six-wheeled rover is packed with equipment to help scientists learn more about Mars and its ability to host life.

Unlike many robots in space, *Curiosity* doesn't need the sun for power. Its energy source is plutonium, an element that creates heat, which the robot turns into power. This lets *Curiosity* travel farther and for longer lengths of time than earlier Mars rovers such as *Spirit* and *Opportunity*.

Curiosity has its own weather station on board.

Though *Curiosity*'s mission was expected to last only about two years, it's still used by NASA today to explore Mars.

Fast Fact

Curiosity has a laser, or strong beam of energy, that can burn holes in rocks up to 23 feet (7 m) away!

MISSION TO MARS

Just how does a rover get all the way to Mars? At its closest, Mars is still 33.9 million miles (54.6 million kilometers) away, and spacecraft have to make it through the harsh conditions of space and new atmospheres. A rover's spacecraft includes a lander, which protects the rover. A part called an aeroshell protects the lander and rover during entry into Mars's atmosphere. Because this entry involves high temperatures, the aeroshell has a heat shield made of special materials. Airbags all around the spacecraft are let out for a softer landing, and a large cloth called a parachute is let out for a slower landing.

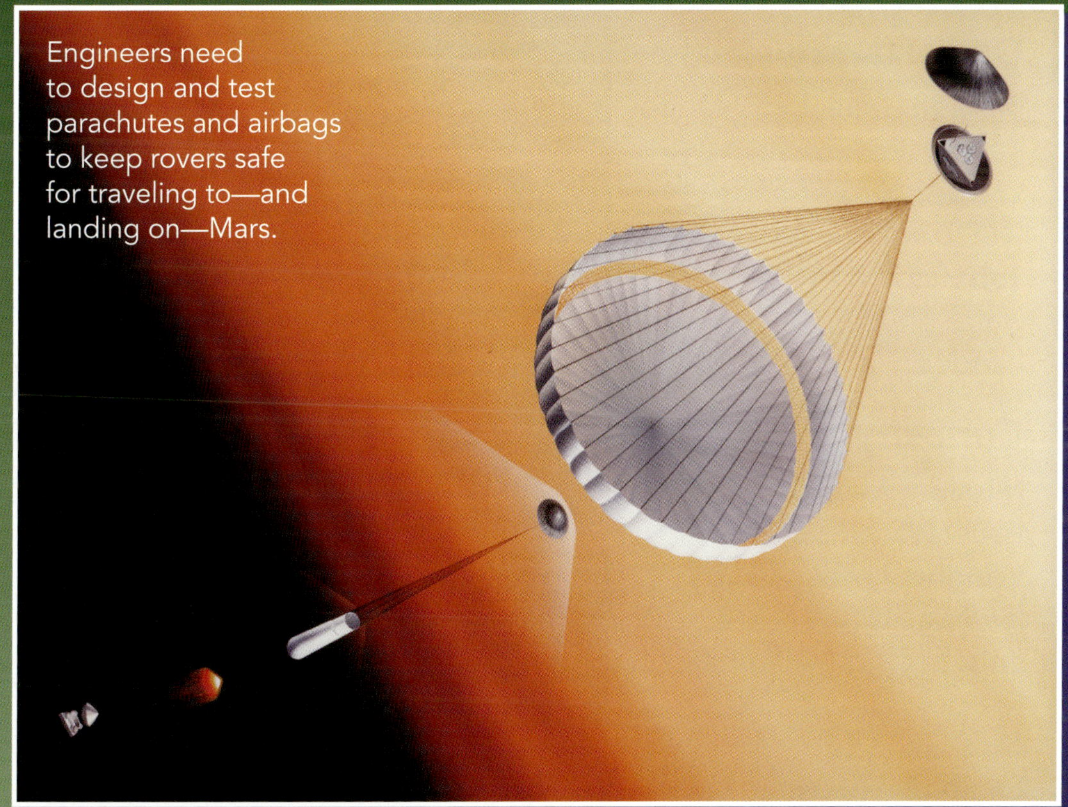
Engineers need to design and test parachutes and airbags to keep rovers safe for traveling to—and landing on—Mars.

Mars Rovers

Sojourner

Mission: Mars Pathfinder
Weight: 23 pounds (10.4 kilograms)
Job: first rover on Mars, took 550 photographs
Landed: July 1997

Spirit and *Opportunity*

Mission: Mars Exploration Rovers
Weight: 374 pounds (169.6 kg)
Job: finding evidence of water on Mars
Landed: January 2004

Curiosity

Mission: Mars Science Laboratory
Weight: 1,982 pounds (899 kg)
Job: finding evidence of the possibility of life on Mars
Landed: August 2012

This keeps track of the wind speed, air pressure, humidity (air moisture), and temperature on Mars's surface. It also has its own robotic arm, which is used to grab rock samples. Its onboard chemistry lab can analyze these samples to learn information, which it sends back to Earth with its antenna. This robot is a master of Mars reconnaissance!

An Amazing Discovery

Rovers have made many important discoveries on Mars. Perhaps one of the greatest is water. Until recently, Earth was thought to be the only planet to have liquid water on its surface. Later, scientists thought Mars had liquid water on its surface at one time but that it was no longer present.

In 2015, *Curiosity* found liquid water just below the surface of Mars. Later that year, the Mars Reconnaissance Orbiter (MRO) discovered recent signs of **intermittent** flowing liquid water on Mars's surface. The MRO can't roam the surface like a rover, but it has six different tools to track potential water on the planet.

This is a model of the design of the Mars Reconnaissance Orbiter.

In 2019, *Curiosity* found large amounts of clay on Mount Sharp on Mars. Clay is direct evidence of water. Learning more about Mars's water history tells scientists about the possibility of life on the Red Planet!

Fast Fact

The MRO used a tool called an imaging spectrometer to track the trails left behind by water flowing on the surface of Mars.

Astronauts have to deal with many challenges while they work aboard the ISS—one of them being low gravity that makes them float!

ASTRONAUT HELPERS

It takes a lot of work to keep humans alive outside of Earth, and robots are necessary helpers on space missions. Robots allow astronauts to travel in space, make observations, and repair spacecraft.

Robonaut on the ISS

Science fiction is full of walking, talking **humanoid** robots, but do they really exist? Robonaut is probably the closest thing NASA has to this kind of robot. NASA began work on Robonaut in 1996. It has a head, upper body, arms, and hands just like a human. Robonaut's head has cameras that allow it to see. The robot works in two different ways: People give Robonaut a simple command to follow, or people use its cameras to see and then command the robot by remote control.

Fast Fact

NASA has hopes that future versions of Robonaut could travel deep into space and help with space walks. Space walks involve astronauts performing a mission outside a spacecraft in space.

Robonaut 2 went into space with a shuttle flight in 2011. Robonaut 2 was able to do simple tasks, such as flipping switches, so astronauts aboard the space station could focus on other tasks. While it originally just had an

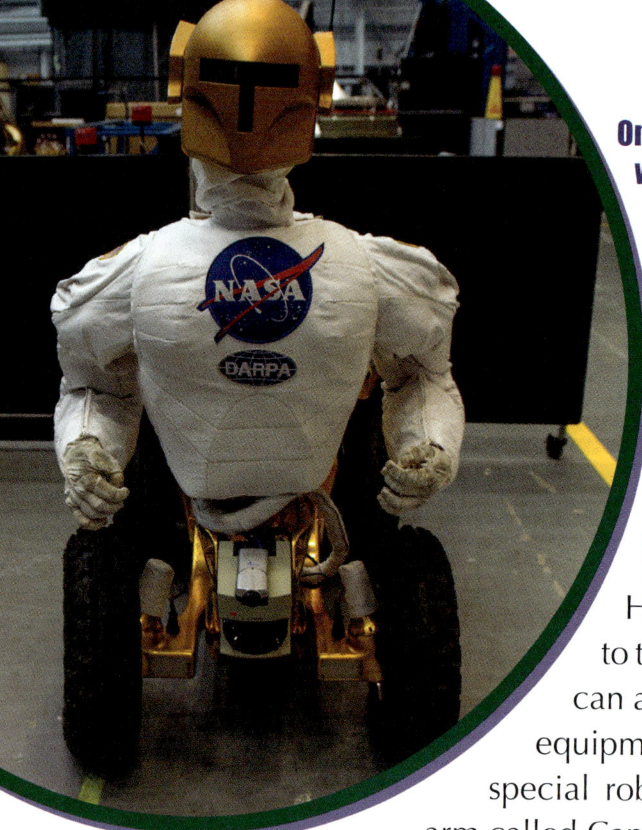

One version of Robonaut was tested with wheels to get around the ISS.

upper body, it was upgraded with legs. Robonaut 2 had some problems, however, and was sent back to Earth to be fixed in 2018.

Robot Repairs

How can astronauts make repairs to the ISS and other spacecraft? How can astronauts pick up huge pieces of equipment in space? They need a very special robot for the job—a huge robotic arm called Canadarm.

The first Canadarm (also known as the Shuttle Remote Manipulator System) was used on a space shuttle in 1981. It was used to grab things outside the shuttle, such as the Hubble Space Telescope and other satellites. It marked the real beginning of advanced space robotics.

Today, the ISS has a larger robotic arm called Canadarm2. This robotic arm is 55.8 feet (17 m) long and has been used since 2001 to make repairs and add

Robonaut 2 was more dexterous, or better able to use its hands to do work, than the first version.

Fast Fact

On April 28, 2001, Canadarm and Canadarm2 worked together to move a huge pallet from the ISS to a shuttle.

Canadarm worked on 90 shuttle missions before it was retired in July 2011.

to the ISS and other spacecraft. The arm can be controlled by an astronaut aboard the ISS or even by someone back on Earth at mission control. Astronauts use a **joystick** to

The ISS also has a smaller, two-armed robot called Dextre that can work with Canadarm2 to make repairs outside the ISS.

A ROBOT HANDSHAKE IN SPACE

Astronauts and engineers had one big challenge to figure out— how to get Canadarm2 set up in space. Since it would be a permanent fixture at the ISS, this huge robotic arm needed to be **installed** correctly. Two astronauts had to train for a space walk to handle the arm once it arrived on April 28, 2001. Once the shuttle with Canadarm2 arrived at the ISS, the original Canadarm was used to lift it from the cargo hold and pair it with the ISS. Astronauts had to then connect and set up the arm. Then, Canadarm and Canadarm2 made history by "shaking hands"—the first robot handshake in space.

control the robotic arm, moving it into place and grabbing things like a human hand could. Unlike the first Canadarm, which returned to Earth after each shuttle mission, Canadarm2 is in space permanently.

A Swarm of Astrobees

Astronauts have a lot of work to do, and sometimes they need a little help. In 2019, astronauts on the ISS were introduced to Astrobee, a free-flying robotic system. In this system, three cube-shaped robots fly through the space station using cameras

Fast Fact

The three Astrobees on board the ISS are named Honey, Queen, and Bumble.

The Astrobee system includes three flying robots, as well as a docking station and software that can be upgraded as needed.

and sensors to navigate. They each have an arm that lets them grasp and hold objects.

Astrobees are designed to move cargo around the station, record experiments, and count supplies. In the future, they may act as caretakers for spacecraft while astronauts are away. Astrobees feature more advanced technology than SPHERES robots, which were robots the size of soccer balls that were brought to the ISS in 2006.

President John F. Kennedy looks at a model of space probe *Mariner 2* in 1963.

EXPLORING THE UNIVERSE

Robots can go much farther in space than humans. It can take years to get to other planets and space objects, but robots don't need food, water, or medical care. They're the perfect explorers of the universe!

A Long Voyage

Space probes were some of the first man-made spacecraft. Probes are robots designed to study things in space. NASA launched the *Mariner 2* probe on August 27, 1962. *Mariner 2* flew by Venus on a three-month mission. It weighed 450 pounds (204 kg) and carried six different scientific tools to study Venus's temperature, magnetism, dust makeup, and more. It had a two-way radio to send data back to Earth and let scientists at the Jet Propulsion Lab in Pasadena, California, track the probe's path through space.

The *Voyager 1* space probe was launched in 1977. It flew past Jupiter and Saturn. In 2013, NASA announced that *Voyager 1* was the first man-made object to

Fast Fact

Mariner 2 learned that Venus's surface is about 900 degrees Fahrenheit (482.2 degrees Celsius) before it lost contact with Earth on January 3, 1963.

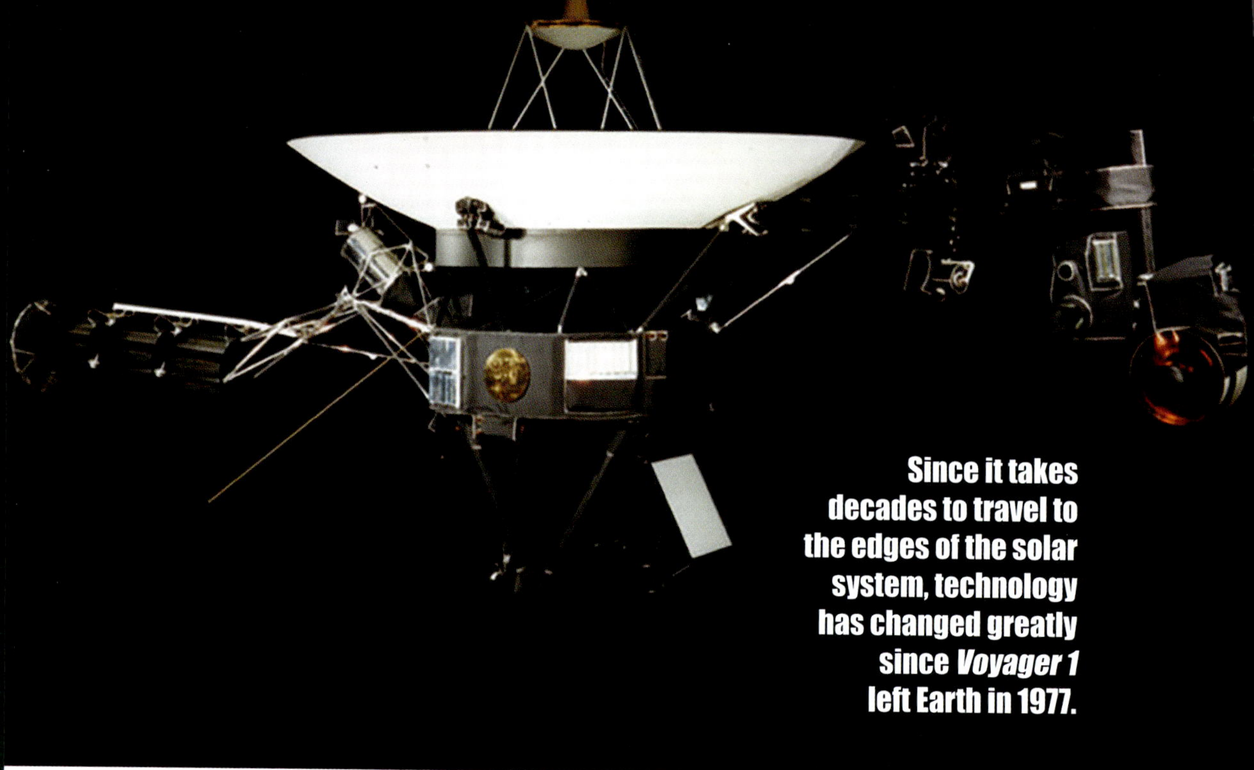

Since it takes decades to travel to the edges of the solar system, technology has changed greatly since *Voyager 1* left Earth in 1977.

reach interstellar space. That means it left our solar system! It has gone farther than any other man-made object.

Cassini Studies Saturn

Saturn is one of the biggest planets in our solar system—nine times wider than Earth. It's also the farthest planet that was discovered without the use of telescopes. There's a lot we still have to learn about this **gas giant**. However, much of what we know about the planet comes from the orbiter *Cassini*.

This very clear photograph of Saturn was taken from the *Cassini* orbiter.

Fast Fact

When *Huygens* landed on the surface of Titan, it became the most distant spacecraft landing ever.

Cassini became the first spacecraft to orbit Saturn on July 1, 2004. *Cassini* carried a probe called *Huygens,* which used a parachute to land on Saturn's largest moon, Titan, in January 2005. *Cassini* and *Huygens* sent back data about Saturn, its moons, and the rings that circle the gas giant. *Cassini* had 12 scientific instruments to study Saturn, including the makeup of its dust particles and its atmospheres. *Cassini*'s sensors were able to "see" light and energy and "feel" magnetic fields.

Robots such as Valkyrie (Robonaut 5) may help NASA explore new locations in space in the future.

Robots of Tomorrow

Robots advance each year, giving humans the opportunity to learn more and more about space. NASA is constantly testing new robots to help astronauts in space and scientists on Earth study the universe. New robots will be able to go farther and may also make it easier for people to live in space.

Robots are the key to astronauts living well on the ISS. For example, some engineers hope a small robotic arm could be used inside the ISS in case of an emergency, such as an astronaut getting sick or hurt. If an astronaut needed surgery, for example, a doctor on Earth could take control of the arm and perform the surgery remotely. Maybe the robots of tomorrow will be able to do all the chores on board the ISS!

In 2005, NASA started the NASA Centennial Challenges program, which allowed people to send ideas to advance space robotics. Prizes

AUTONOMOUS ROBOTS

Autonomous robots are robots that can perform tasks on their own, without the control of a human. These robots use artificial intelligence to decide which behaviors will complete their jobs. NASA wants to develop autonomous robots to help future astronauts on long missions to faraway places. They believe these robots will help astronauts explore the surface of the moon and Mars. One robot design is the Robonaut 5 (also called R5, or Valkyrie), a humanoid robot that can assist a human mission to the moon or Mars. The R5 could assemble equipment before the astronauts arrive and do repairs during the mission as needed.

Can you imagine a robot making it all the way to another planet or solar system? Robotics engineers can make it happen.

are awarded to individuals, companies, and even student groups for **innovative** ideas. Do you have an idea for a space robot? Would you like to go to space someday? Maybe your ideas today can lead to the space robots of tomorrow!

THINK ABOUT IT!

1. Some space robots are humanoid. In which cases do you think a humanoid robot would be most helpful?

2. The first three Mars rovers were powered by solar panels. What do you think are the benefits and challenges of powering a robot with energy from the sun?

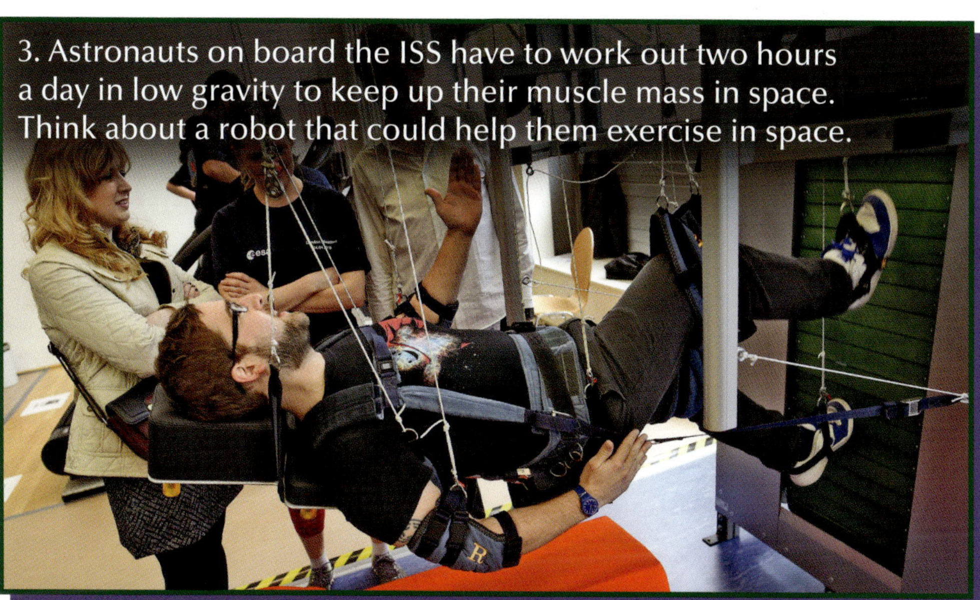
3. Astronauts on board the ISS have to work out two hours a day in low gravity to keep up their muscle mass in space. Think about a robot that could help them exercise in space.

4. Autonomous robots can travel and work without control from a human operator. What jobs would you assign to an autonomous robot in space?

GLOSSARY

artificial intelligence: An area of computer science that deals with giving machines the ability to mimic intelligent human behavior.

atmosphere: The mixture of gases that surround a planet.

gas giant: A large planet of low density, mostly made up of gases, such as Saturn, Jupiter, Uranus, and Neptune.

humanoid: Resembling a human.

innovative: Using or showing new methods or ideas.

install: To set up for use or service.

intermittent: Coming and going; not continuous.

joystick: A control device that allows motion in two or more directions.

lens: A clear, curved piece of glass or plastic that changes the direction of light rays.

orbit: To travel in a circle or oval around something, or the path used to make that trip.

repetitive: Occurring over and over again.

satellite: An object that circles a planet in order to collect and send information or aid in communication.

FIND OUT MORE

Books

Colins, Luke. *Space Robots*. Mankato, MN: Black Rabbit Books, 2020.

Larson, Kirsten W. *Space Robots*. Mankato, MN: Amicus Ink, 2018.

Smibert, Angie. *Space Robots*. Minneapolis, MN: Abdo Publishing, 2019.

Websites

***Curiosity* Rover**
www.nasa.gov/mission_pages/msl/index.html
Follow *Curiosity*'s mission updates as this rover makes its way around Mars.

Hubble Captures Colorful Universe
kids.nationalgeographic.com/explore/space/hubble-colorful-universe/
Explore a slideshow of space photographs from the Hubble Space Telescope.

The Mars Rovers: *Perseverance*
spaceplace.nasa.gov/mars-2020/en/
Learn all about the newest Mars rover on a mission—*Perseverance*!

INDEX